NATURE'S MAGIC
CAMOUFLAGERS

WRITTEN BY
Q. L. Pearce

ILLUSTRATED BY
Marilee Niehaus

To Brittany Schmidt
—Q.L.P.

To Curtis Lee and my grandfather, Sam
—M.T.N.

PRICE STERN SLOAN
Los Angeles

Copyright © 1991 by RGA Publishing Group, Inc.
Published by Price Stern Sloan, Inc.
Los Angeles, California

Printed in the United States of America. All rights reserved.
No part of this publication may be reproduced, stored in a retrieval system or transmitted, in any form or by any means, electronic, mechanical, photocopying, recording or otherwise, without the prior written permission of the publisher.
10 9 8 7 6 5 4 3 2 1   ISBN: 0-8431-2828-3

Designed by Heidi Frieder

THE KATYDID

A great impostor

When is a leaf not a leaf? When it's an insect! Many animals have developed protective colorings to avoid being gobbled up by other hungry creatures. The slender katydid has taken this strategy one step further. Not only is it the *color* of a leaf, but it looks as though it really *could* be a leaf. This type of camouflage is called "special resemblance."

Some ground-dwelling katydids are brown. They look like dry, dead leaves, complete with light spots that appear to be growths of fungus. Other katydids are green, with patches that resemble the damage done by a chewing insect. These katydids even have light veins on their wings that resemble the veins of a genuine leaf. The clever katydid has another trick, too: When the wind rustles the leaves where a katydid is hiding, the clever counterfeit moves just a little, as if it, too, were a leaf fluttering in the breeze.

THE TIGER

A camouflaged cat

The tiger's striking coloration of orange and black stripes is actually a form of camouflage. Of all the great cats, such as lions, leopards and cheetahs, the tiger is the largest. Like its relatives, it is a fine hunter. A forest dweller, the tiger hunts at night, relying on cunning to capture its dinner. Up to 9 feet long and weighing up to

600 pounds, this giant has a huge appetite, too. It can gobble down ninety pounds of meat in one meal. The magnificent tiger is powerful but not particularly fast, so it needs to be able to creep through the tall grass unnoticed as it tries to get closer and closer to its prey. Under the cover of darkness, it lurks in ambush for a deer or goat. When it discovers a likely target, the mighty cat slowly edges silently forward, preparing to pounce. In the gloomy forest, its striped coat looks like a pattern of shadow and light. The prey animal is often unaware of the dangerous cat until it is too late.

THE CHAMELEON

A quick-change artist

The word *chameleon* means "changeable," and it suits this lizard to a tee. Special cells in this slow-moving reptile's grainy skin contain pigment, or color. When the arrangement of the pigment is altered, the animal rapidly changes hue. In bright sunlight, it may become a light, sunny yellow-green. In the cool shadows among the leaves, the animal darkens to a deep green or gray-brown.

This ability to change hue gives the chameleon an advantage in stalking a meal of insects. Surprisingly, though, disguise is not the main reason for the color change. The transformation is actually a response of the lizard's nervous system to light, temperature or emotions (such as fear). In changing color, the chameleon may be warning its rivals or attracting a mate.

THE HUNTSMAN SPIDER

Bark with a bite

American Indian hunters of long ago camouflaged themselves with animal skins so they could creep very close to herds of bison without alarming them. Modern hunters camouflage themselves differently. To blend in with their surroundings, they wear colored and patterned clothing. In this

way, they have an advantage over their prey. The huntsman spider of Australia, a skilled hunter of the animal world, relies on a similar tactic.

The territory of the huntsman spider is tree bark. It eats insects that march across the trunk and burrow into the wood. At four inches across, this spider would be easily detected if it prowled about looking for a meal. Instead, the patient spider sits motionless, waiting for its food to pass nearby. Its dappled gray and brown body matches the bark of the tree, so it goes unnoticed by its unsuspecting prey.

THE WHIP-POOR-WILL

A voice in the night

The whip-poor-will is a bird that is heard but not seen. When darkness falls, this member of the nightjar family calls what sounds like its own name over and over in a loud, insistent tone. Listeners once heard a whip-poor-will call its name more than 400 times in a row!

About the size of a pigeon, the whip-poor-will lives in dry, open woodlands in the eastern United States. It usually sleeps during the day on the leaf-strewn forest floor. That may not sound like a safe place for a bird, but the whip-poor-will is protected by its coloring. Ash-gray, beige and brown feathers make it look like a pile of dried leaves and twigs, and patches of light and dark mimic the play of light and shadow among the leaves. The whip-poor-will even lays its eggs on the ground. Streaked with gray and brown, they, too, are cleverly hidden in plain sight.

THE PEPPERED MOTH

A town-and-country insect

The amazing peppered moth is an example of not only the success of camouflage, but of the incredible adaptability of animals. These insects make tasty morsels for hungry birds. A moth that stands out is likely to be eaten, but a moth that can blend with its background while resting is more likely to be overlooked.

In the English countryside, the trees are speckled with dark, plantlike growths called lichens. The wings of the peppered moths that live in the country match the speckled tree bark, helping the insect to survive. In cities, smoke from factories and homes often fills the air. Trees have turned charcoal gray and black from the pollution. In this kind of dark environment, the speckled peppered moth would be an easy target. Thus, the wings of city peppered moths are dark to match the sooty tree bark!

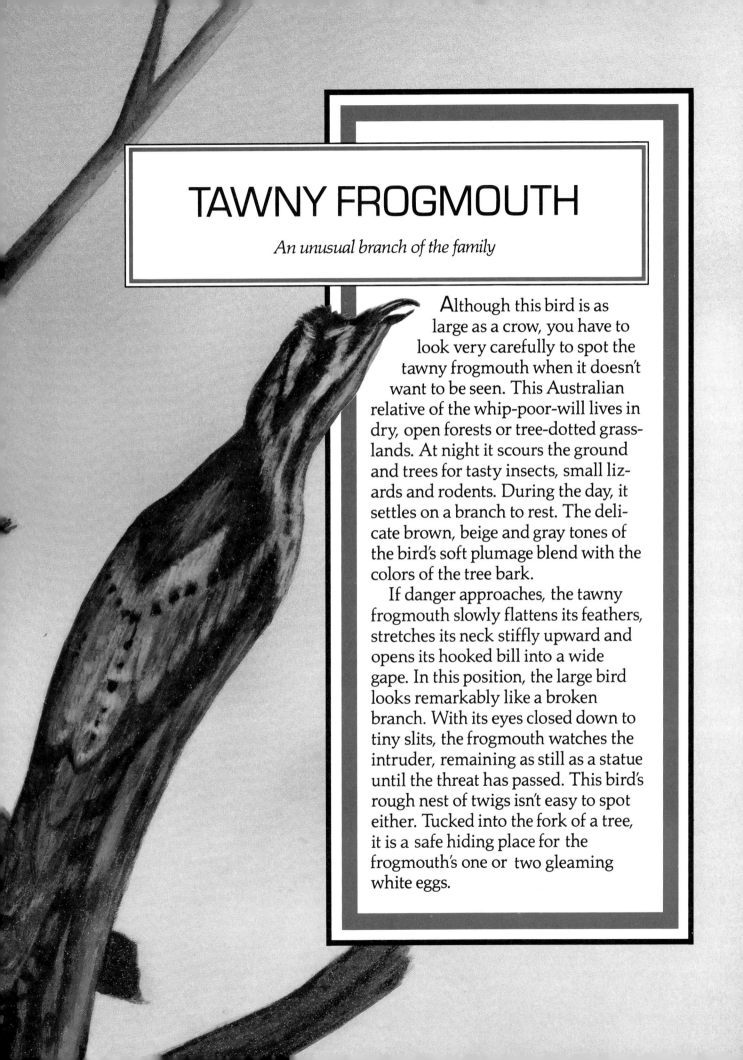

TAWNY FROGMOUTH
An unusual branch of the family

Although this bird is as large as a crow, you have to look very carefully to spot the tawny frogmouth when it doesn't want to be seen. This Australian relative of the whip-poor-will lives in dry, open forests or tree-dotted grasslands. At night it scours the ground and trees for tasty insects, small lizards and rodents. During the day, it settles on a branch to rest. The delicate brown, beige and gray tones of the bird's soft plumage blend with the colors of the tree bark.

If danger approaches, the tawny frogmouth slowly flattens its feathers, stretches its neck stiffly upward and opens its hooked bill into a wide gape. In this position, the large bird looks remarkably like a broken branch. With its eyes closed down to tiny slits, the frogmouth watches the intruder, remaining as still as a statue until the threat has passed. This bird's rough nest of twigs isn't easy to spot either. Tucked into the fork of a tree, it is a safe hiding place for the frogmouth's one or two gleaming white eggs.

TEXAS HORNED LIZARD

A well-prepared reptile

The Texas horned lizard has a bagful of tricks. Camouflage is one of them. This reptile makes its home in the sandy deserts of the American Southwest. Tiny spines along its sides give it a ragged appearance, and its scaly skin blends well with the grainy desert sand. But

even a cleverly disguised creature can be betrayed by its own shadow. So, when alarmed, this lizard flattens itself against the ground so it has no shadow at all. It then wiggles sideways until it is covered with sand.

If these camouflage methods don't succeed, the horned lizard may try a different approach.

To look larger, it inflates its body with air. If the attacker persists, it may be in for a surprise. When the lizard is harassed, a tiny sinus in the corner of each eye may rupture, shooting out a small stream of blood. The blood travels up to three feet, startling the enemy.

THE FLOUNDER

A fish that shows its best side

Swimming near the surface of the sea, a newborn flounder seems to be a rather ordinary fish. But when it is an adult, nothing about this fish seems ordinary. When the flounder is about two weeks old, the bridge of its nose begins to dissolve, and one eye slowly moves to join the other. The fish settles to the ocean floor with its blind side down, and there it remains for the rest of its life. It swims on its side just above the sea floor.

To lie in wait for prey or to hide from predators, this fish has an unusual strategy: It buries itself on the sea bottom. With a quick, shivering movement, the flounder covers its body with sand, leaving only its eyes and mouth uncovered. This fish can also change its appearance to match the sea bed, though this takes more time than burying itself. By expanding or contracting tiny pigment (color) cells in its scaly skin, the remarkable flounder can imitate the pattern of the surface on which it rests.

THE POLAR BEAR

A bear of a different color

The polar bear, an inhabitant of the icy Arctic, has few enemies from which to hide. Instead of serving as protection, its excellent camouflage is a hunting tool, helping the bear to hide from its prey. Each colorless hair of the polar bear's thick coat reflects visible light, making the bear appear white. Thus, the 8-foot-long, 1,500-pound

creature of the north blends perfectly with the snow and ice.
 Wide bands of hair grow between the small footpads on the bottom of the polar bear's feet. The hair not only provides better footing on ice, but it also helps to muffle the bear's footsteps. With its snowy coloring and silent steps, a polar bear can creep very close to the seals that it hunts. Sometimes it hides behind snow drifts to get near a resting seal, then bounds across open space to reach its prey. More often, the giant bear crouches by a breathing hole in the ice and waits for a seal to surface for a breath—
and be captured!

THE SPINY SPIDER CRAB

An exterior designer

Many camouflagers are born with natural disguises, but the spiny spider crab designs its own clever disguise bit by bit. Since this sea creature is generally a scavenger (which means it feeds on animals that are already dead), it doesn't need its camouflage to hunt. Rather, it needs to hide for its own protection. This slow-moving crab may itself fall victim to predators—if they can find it, that is!

The spiny spider crab's hard upper shell is covered with short, prickly spines. Using its strong pincers, this resourceful, six-inch-wide crab clips and tears up tiny clumps of seaweed. It then dabs the clumps with a sticky, mucuslike substance from its mouth and sticks the clumps onto its spines. By adding seashells or bits of sponge to its disguise, the spiny spider crab can blend right into its background. If it moves to a different area, it can readily "remodel" its shell with local seaweed.

SCORPION FISH

Danger in disguise

Members of the venomous scorpion fish family rely on camouflage to blend with their undersea background of coral, seaweed or rocks. This helps them to capture prey. More important, it also helps them to avoid becoming a meal themselves!

The most dangerous of the scorpion fish, the stonefish masquerades as a seaweed-covered rock. Its mottled appearance makes this deadly animal almost invisible among the rocks as it lies on the shallow sea bottom. When it is startled, the stonefish raises sharp spines that are connected to venom glands. If a wading human accidentally steps on the stonefish, the animal's spines easily pierce the person's foot and inject a powerful poison into the wound.

THE HORNED FROG

A hunter in hiding

The South American jungle contains more forms of animal life than any other place on Earth. In the treetops, hundreds of varieties of birds, monkeys and insects flutter, scamper, chatter or buzz. The jungle floor is densely populated too, but many of the ground-dwelling creatures are hard to see. The frog-eating horned frog

is just such an animal.

This amphibian squats still and silent amid the damp leaf litter. A patchy pattern of light and dark brown on its skin makes the horned frog appear to be nothing more than a pile of dead leaves. Its dark eyes are like shadows under the little pointed ridges, or horns, that project over them. In the dangerous world of the jungle floor, the horned frog uses its camouflage to escape notice by the small frogs it eats, as well as by the snakes and mammals that would like to eat *it*! If by chance the frog is disturbed, it can deliver a painful bite.

THE THREE-TOED SLOTH

A treetop loafer

As you read this, a three-toed sloth is this very minute hanging in a tree in the Amazon rain forest doing nothing, and doing it well. This animal sleeps high above the forest floor for up to twenty hours a day, gripping a branch with its long, curved claws. Remaining still for hours on end is part of the sloth's strategy for survival.

Rain often pours down on this peaceful creature as it slumbers. Because the sloth hangs belly up, rainwater runs easily through the hair, leaving the animal's coat damp. The moist fur makes a perfect place for tiny plants called algae to grow. The algae tint the sloth's fur pale green, making the creature nearly impossible to see among the dense leafy branches of its rain forest home.

MALAYSIAN FLOWER MANTIS

A beastly beauty

The praying mantis is so named because it seems to be holding its forelegs up in prayer. But the truth is that this predatory insect could more properly be called a "prey-ing" mantis. Like all mantises, it is actually holding its forelegs up in preparation for attack. If a tasty insect crawls within range, the mantis's forelimbs snap outward at incredible speed to grasp the prey.

The limbs then snap back, trapping the unfortunate creature in a viselike grip.

One kind of mantis—the ivory-white or petal-pink Malaysian flower mantis—is so beautiful you wouldn't think it was a cunning hunter. This native of the Southeast Asian rain forest could easily be mistaken for a delicate blossom of the plant on which it lurks. But the mantis doesn't hide. In fact, it often turns toward the sun to call attention to itself. Insects are lured to what they think is a flower, hoping to feed on sweet nectar, but they soon find themselves on the menu.